CONTENTS

Precision-Guided Weapons

INTRODUCTION

Ever since military men began shooting things at enemies, most shots have missed or been ineffective. The remarkable thing that has happened over the past few years is that new weapons have been developed which can hit with most of their shots, usually effectively. While much of the theory of these weapons is not new, it is only since laser-guided bombs were used in Vietnam that military planners have generally agreed that they were economically and operationally feasible. The early name 'smart bomb' was dropped, and a larger class of 'precision-guided munitions' (PGM) became official. Usually this term simply refers to a bomb or missile that is guided during its terminal phase, thus including many anti-tank weapons and air-defence missiles. It was to these that attention shifted after their widespread use during the October war of 1973, especially by Arab forces.

In the first part of this Paper I shall say enough about the mechanics of these weapons to give the reader a feel for how they work, and provide brief descriptions of some of the more important weapons associated with non-nuclear land combat. However, the brief treatment here mentions only a fraction of current PGM developments (it is characteristic of the pace of development that dozens of new PGM types in many countries reach the operational testing stage each year). The second part of the Paper focuses on a number of important and so far unresolved implications of the new weapons, and discusses their likely effects on force posture and the conduct of warfare. For example, what is their effect on the usefulness of the advanced tank, the complex fighter-bomber and the big aircraft carrier? What will be the consequences for the organization of land forces and for their tactics? Land forces may need to adopt a kind of molecular posture of many highly mobile and concealable – but powerful – squads, and it may be that a smaller weight of munitions will have to be hauled to the battle area. What are the political consequences? If barrage fire and carpet bombing are not needed there may be less collateral damage to civil populations and the economy, and there are prospects for raising the threshold at which nuclear weapons would be used (the consequences of this are both encouraging and urgently in need of study). On the positive side (from an American point of view) there is a likelihood that the resulting postures will be advantageous to NATO and to American strategies.

Serious consideration of many of these points has only just begun, and the devising of counters to PGM is just gaining momentum. Thus the reader should regard much of what follows as tentative. After all, military analysts foresaw only dimly the implications of some very predictable technologies, and it took the 1972 bombing raids in Vietnam and the October 1973 war in the Middle East to get serious exploitation of PGM under way.

It is beyond the scope of this Paper to say very much about technological trends or to explore adequately countermeasures and counter countermeasures.

I. EXAMPLES OF PGM

This part describes several precision-guided munitions – of different kinds – to give the reader a better idea of the scope of what will be discussed. A pedantic definition of a PGM would run something like: 'A guided munition whose probability of making a direct hit at full range – when unopposed – on a tank, ship, radar, bridge or aircraft (according to its type) is greater than a half.'

ADELPHI PAPERS
NUMBER ONE HUNDRED AND EIGHTEEN

Precision-Guided Weapons

by James Digby

THE INTERNATIONAL INSTITUTE FOR STRATEGIC STUDIES
18 ADAM STREET LONDON WC2N 6AL

ADELPHI PAPER NO. 118

James Digby is a senior staff member at the Rand Corporation. The views expressed in this Paper, however, are those of the author. They should not be interpreted as reflecting the views of the Rand Corporation or the official opinion or policy of any of its governmental or private research sponsors.

This Paper draws on an earlier Paper presented at the conference on 'The Impact of New Technologies on US Defense Planning for Non-nuclear Conflict', sponsored by the Fletcher School of Law and Diplomacy, Tufts University, September 1974.

First published Summer 1975

ISBN 0 900492 87 2
ISSN 0567-932X

The International Institute for Strategic Studies was founded in 1958 as a centre for the provision of information on and research into the problems of international security, defence and arms control in the nuclear age. It is international in its Council and staff, and its membership is drawn from over fifty countries. It is independent of governments and is not the advocate of any particular interest.

The Institute is concerned with strategic questions – not just with the military aspects of security but with the social and economic sources and political and moral implications of the use and existence of armed force: in other words, with the basic problems of peace.

The Institute's publications are intended for a much wider audience than its own membership and are available to the general public on special subscription terms or singly.

Many discussions exclude surface-to-air missiles, principally because they have been quite precise for a long time; what is new in their case is being precise *and* relatively cheap *and* easy to operate. But note the adverb relatively. Many of the new weapons are quite sophisticated and can only be called cheap in relation to earlier guided missiles or to the destructive capacity of other weapons that might be used for similar jobs. However, they are not too expensive to preclude an abundant supply – and the possibility of abundance accounts for much of the significance of PGM.

General awareness that something new had happened was triggered by the performance of the unpowered *Pave Way* laser-guided bombs in Vietnam in 1972. With fairly simple aids, a narrow laser beam from a 'designator' was pointed at the target from an aircraft. A two-part kit attached to Mk 82 500-lb or Mk 84 2,000-lb bombs provided steerable front fins (controlled by a laser receiver) that homed on the energy reflected from the target. In earlier systems the laser beam was aimed at the target from a spotter aircraft, but a later version avoided mix-ups by designating from the F-4 bomb-carrier itself. Excellent accuracies could be obtained, making it possible to destroy in one or two sorties a bridge span that might otherwise have required dozens. Judging from the aerial photographs released by the US Air Force, the accuracy equalled or exceeded the requirements set in 1968: 'CEP no greater than 25ft; guidance reliability at least 80 per cent.'[1]

The Soviet *Sagger* AT-3 wire-guided anti-tank missile saw extensive use in the October 1973 war in the Middle East. Often mounted in sixes or eights under a kind of steel umbrella on the BRDM-2 armoured car, it weighs 11 kg, has a 2·7kg warhead and takes 25 seconds to reach its maximum range of 3,000m (which is long enough to allow the target to seek cover or to distract the guider by taking him under fire). Similar missiles on the NATO side are the American *TOW* and the Franco-German *HOT*, both of which are faster and have semi-automatic tracking features that mean the gunner need only track the target, rather than having to 'fly' the missile into the target. The British *Swingfire* is less automatic, but its guidance unit can be positioned up to 100m away from the launcher. All three have helicopter-mounted versions.[2]

Let us examine in a little more detail how one of these anti-tank missiles works, taking *TOW* BGM-71 for our example. The missile is 15cm in diameter and weights 23·6kg including its protective container, which serves as a launch tube. The container is sealed at the factory, and the missile is used without further checking. The *TOW* launcher is mounted on a tripod which can be used independently of a vehicle, or, very commonly, on a metal column fastened to the M-113 armoured personnel carrier. The gunner sights on the target vehicle through a 13-power monocular telescope; if there is a choice of targets, he selects the one most likely to stay in sight during the time of flight of the missile. When he presses the firing button the rocket motor ignites, leaving a signature significantly less than that of the 106mm recoilless rifle which *TOW* has replaced. As the missile flies towards the target, control wires attached to the launcher unwind from two bobbins, each about 18cm long, mounted in the tail of the missile. A modulated infrared source in the missile tail is then followed by an infrared tracker mounted in the launcher, and steering commands, based on the difference between the angles from launcher to missile and from launcher to target, cause the missile to move its aerodynamic fins, and direct it to fly a course roughly along the line of sight to the target. Meanwhile, a safety and arming device arms the warhead, a 3·6kg shaped-charge device,[3] which detonates on contact.[4]

In training, *TOW* can engage three targets within a 90-degree angle in less than 60 seconds, though some experienced officers doubt if such a rate would usually be possible in combat. The 1974 production version of *TOW* has a range of 3km and a time of flight over that distance of about 15 seconds. An extended-range version

[1] Quoted in P. G. DeLeon, *The Laser-Guided Bomb: Case History of a Development* (Santa Monica, Calif.: Rand Corporation, R-1312-1-PR, May 1974). This report includes a useful bibliography.

[2] Data from *Flight International*, 14 March 1974.

[3] A shaped charge (or hollow charge) has its explosive so shaped as to focus the blast wave generally along the axis of flight. The hot concentrated gases can burn a hole in armour which might withstand a non-directional blast from the same weight of explosive.

[4] *Jane's Weapon Systems, 1973–74* (London: Jane's Yearbooks, 1973).

has been developed with a maximum range of 3·75km. At present the cost of the missile is approximately $3,000, while the infantry-version guidance unit costs about $20,000.[5]

Maverick AGM-65A, an air-launched anti-tank missile developed by the US Air Force, is steered by a stabilized television camera in the nose which feeds video signals into automatic tracking circuits in the missile; these generate guidance signals. Six missiles can be carried on F-4, A-7, and A-10 aircraft. The pilot manoeuvres to get his sight lined up with the target, and when he does so servo controls lock the missile's automatic tracking device on to the same target. When the cross-hairs of the missile's target-seeker frame the target, a lock-on switch is actuated and the missile's contrast-seeker automatically tracks the target. The missile is then launched. At any time after launching, which can take place several miles from the target, the aircraft can leave, or the crew can repeat the procedure and attack other targets. *Maverick* is 30cm in diameter, carries a relatively heavy warhead and has a total weight of 210kg. For the quantity procurement begun in 1974 the unit price was under $10,000. A laser-guided version and an infrared version are under development.[6]

American forces first saw the Soviet *Grail* SA-7 anti-aircraft missile used in action in Vietnam.[7] It can be carried by an infantryman and launched from the shoulder. It has an infrared seeker and a 1·1-kg warhead. Reports on the October war indicate that it was fired in salvo at single Israeli aircraft, 'damaging the jetpipes of many Israeli A-4s but not achieving a very high kill ratio'.[8] However, the Egyptian forces which used the SA-7 probably proved its operability. The SA-7 is similar to the current American *Redeye*, which is being replaced by an improved version called *Stinger*. The British *Blowpipe* is an optically guided missile of similar size, also fired from the shoulder.

Standard ARM (anti-radiation missile) AGM-78 has been developed by General Dynamics for use by US Navy and US Air Force attack aircraft against missile radars and other radiating targets. The seeker design is based on that of the *Shrike*, which saw extensive use in Vietnam against surface-to-air missile radars. A signal-intercept unit aboard the aircraft processes received signals and gives the missile homer the necessary data to permit it to lock on. After launch a passive radiation seeker steers the missile;[9] if the hostile radiation source shuts down, stored information is used to guide it on a less accurate trajectory. Anti-radiation missiles, unlike all the PGM described above, are fully operable in darkness and bad visibility and are particularly useful for suppressing such emitting targets as SAM radars. This growing class of weapons is likely to be much more widely used in the future against other emitting targets, such as command and control centres.

Consideration of two hypothetical missiles of somewhat larger size will round off this list of typical PGM. While components for missiles like these are in development, and missiles even have been tested as complete systems, no fully operational weapons like these exist in quantity.

First, there is the possibility of a 100-km range air-launched cruise missile that gets its mid-course guidance from ground-based transmitters, then, for the last 10km, it corrects its course to target by means of an optical area correlator. This is a device whose electronic circuits compare a map-like picture of the terrain below (received by an on-board sensor) with a reference picture taken on a reconnaissance flight – probably from very high altitude. This kind of missile would be well-suited to attacking a fixed target, such as a depot or airbase, and its map-matching terminal guidance would be very hard to jam. It could, with a well-organized operation, be used against sporadically moving targets.

Second, let us consider a possible alternative way of doing the same job. This could be an air-launched remotely piloted vehicle (RPV) of about the same size and range, guided by a pilot who watches a television picture relayed from the RPV and sends his steering and throttle signals by radio. In addition to attacking fixed targets, this missile could be used against moving or

[5] *Ibid.*, and *TOW Weapon System Status* (Hughes Aircraft Company, December 1973).

[6] *Flight International*, *op. cit.*, p. A1.

[7] The Soviet Union does not normally give official nicknames to such weapons; *Grail* is the NATO nickname. Journalists in Indochina reported the name as *Strella*, which may have been a phonetic spelling, via Vietnamese, of the Russian *strela* (arrow).

[8] *Flight International*, *op. cit.*, p. A12.

[9] A 'passive' seeker receives signals but does not transmit; an 'active' seeker has both transmitter and receiver.

movable targets. The US Navy's *Condor* AGM-53A already has these properties, among others, but refinements in guidance and anti-jamming features have driven its price up to more than $200,000; a simpler design, more like the Advanced Research Project Agency's experimental *Praeire*, a remotely piloted model aircraft of lower performance, may be able to do many of the same jobs.

The foregoing is enough to indicate that the term PGM covers a broad class of guided bombs and missiles. RPV are usually included if they are intended to hit a target (these are sometimes called 'kamikaze RPV'.)

II. IMPLICATIONS OF PGM TWENTY YEARS HENCE

Let us now turn from what PGM are to some speculations on the consequences of having them. I will use a kind of funnel approach, taking the broader and longer-term consequences before coming to the nearer future and to some specifics relating to NATO. By discussing PGM in the context of operations in the 1990s we will not be held back by practical and bureaucratic constraints, as we will be in discussing implications for the earlier period.

The basic point about precision-guided weapons can be stated thus:

> Accuracy is no longer a strong function of range, and if a target can be acquired it can usually be hit. For many targets, hitting is equivalent to destroying.

A second point may be equally important:

> Precision-guided munitions can now be mass-produced in great quantity; for many the cost per round ranges from around $1,000 to around $10,000. Moreover, many can be operated by average soldiers.

A number of important propositions follow logically from these statements. I will outline seven of them in simple terms before discussing some complications and the degree to which the simple ideas are applicable in the practical world.

Proposition 1. It will become much less desirable to concentrate a great deal of military value in one place or in one vehicle. This will be especially true where a great deal of value can be destroyed by a single warhead. For instance, a combatant would be less likely to want to place a large fraction of his capability at risk by exposing a single transport aircraft, or a single surface vessel in the Mediterranean, and he would probably prefer to have many inexpensive lightly armoured vehicles rather than fewer more expensive tanks. If the attacker has a finite number of PGM, any one of which has a high probability of destroying its target, then it is better to force him to spread them over many targets which are individually of small value.

Proposition 2. With PGM, seeing a target can usually lead to its destruction. Concentrations of vehicles or men – usually easier to see and keep track of than larger numbers of independently moving targets – will be less practical, and concealment will become more important. Smallness and mobility will make hiding easier, and both these qualities are consistent with the thrust of Proposition 1.

However, one must also consider the degree to which concentrations can still be sheltered, or protected by active defences. It is a classic offensive tactic to attempt overwhelming superiority in a narrow sector by concentrating forces there, and the ability to organize and defend such concentrations is an important factor in any assessment of the balance between offence and defence. The availability of tactical nuclear weapons did not in practice result in fully corresponding action to decrease vulnerability – perhaps because of uncertainty about whether they would in fact be used. However, there is no question of PGM not being used if fighting takes place, and no tactical planner can any longer afford to ignore their effect on his vulnerabilities.

Proposition 3. Even small units can be very powerful when equipped with PGM or with designators that can call in and guide remotely-launched PGM – and they might carry air-defence weapons as well. In land warfare the natural size of many such independently mobile squads might be three or four men, moving on their feet or in inexpensive vehicles, rather than in expensive tanks.[10]

[10] Over the past two years T. F. Burke of The Rand Corporation has developed this and a number of related ideas and discussed them in lectures at the Army War College and other service schools; no published version is available.

Taking the European case, let us suppose that NATO forces were the first to change appreciably towards such numerous, dispersed, concealable, independently mobile front-line units. This would help to reduce vulnerability to both PGM and nuclear weapons, and so would make sense as a unilateral move – so long as the ability of these units to deal with a Warsaw Pact armoured thrust is not seriously degraded. They would also need protection from conventional over-running attacks by infantry, but their mobility and their ability to call in PGM firepower – especially PGM with area coverage – would help. These would seem to be good reasons, however, for both sides eventually to move increasingly towards a kind of molecular posture for forward units. One can speculate that the forward edge of the battle area[11] would thus become even fuzzier, and, in time, these units would have fewer important targets to designate (the pressures to reach for more distant, rear-echelon targets will be discussed later).

Proposition 4. Where forward units serve as spotters and designators, not all the munitions used need to be hauled all the way to the forward edge of the battle area; some might be ground- or air-launched from tens of kilometres further back. Furthermore, in many types of conflict the higher hit probability of PGM means that, to achieve a given effect on enemy forces, the weight of munitions delivered to the launch point need not be nearly as great as in the past. Nonetheless, the value of munitions used per day may be very large (this will be discussed later). One must consider the changes in both total needs and rate of use before one can understand the implications for the size of support elements and the vulnerability of supply lines.

Proposition 5. A natural consequence of having their high hit probability is that PGM are likely to cause much less collateral damage to civilian populations and economies. In the NATO case, this prospect may in the long term have substantial consequences for West Germany's attitude towards preparations for actual fighting on her territory and the damage it might involve. However, this change of attitude would be much less likely if precision-guided sub-kiloton 'mini-nukes' were under consideration (this, in addition to the disadvantages of mini-nukes in blurring the firebreak, is mentioned below on p. 11).

Proposition 6. Ground-based anti-aircraft defences will become extremely lethal. The Soviet SA-7 is a step towards a potentially powerful weapon for the air defence of mobile units, and proves the operational feasibility of this class of weapons; however, as already implied, it seems to be under-designed in warhead, range and speed, even against present-generation aircraft. These deficiencies must have become apparent to the Soviet Union during the October war, and correcting them should be a routine matter. In any event, air defences derived from systems like the ZSU-23-4 four-barrel gun, the SA-6 mobile missile and the SA-7 are likely to proliferate over the area occupied by ground troops. The lighter of these classes of weapon may well be added to the mobile squads mentioned above, and the heavier may travel with them, along with the anti-tank weapons. The result may be a shift in methods of protecting ground forces against enemy aircraft: more protection is likely to be provided by ground-based anti-aircraft defences, and less by air-to-air duels and attacks on enemy air bases.[12]

One point in favour of such anti-aircraft defences is the high 'contrast' (in terms of both visibility and energy emission) of an aircraft against a relatively blank sky; this facilitates target acquisition and homing by PGM. Another is that a multi-purpose fighter aircraft costs around $10,000,000 while the weapon to shoot it down with costs less than $10,000. These ground defences may not have a kill probability even as high as 5 per cent for the area defended by any one weapon, but flying over many defended areas will be very costly.

Proposition 7. Finally, the properties of these new weapons may well lead to a major revision of the assignment of roles and missions to the different services. It is no longer very important what form of transportation carries a munition to the place where it is launched; it gets its

[11] This term is used so as to avoid the implication of shallow depth in 'front line'.

[12] NATO's response to the substantial build-up of Warsaw Pact short-range anti-aircraft weapons has been an emphasis on stand-off air-delivered close-support weapons and on defence suppression. Against aircraft so armed, air-to-air combat is still important. This response – logical, given present NATO aircraft inventories – will in the long run have to compete in cost-effectiveness with close support by RPV, ground- or air-launched from well inside friendly territory.

effectiveness from its warhead and its terminal guidance. This makes it more logical for forces to be organized in terms of the type of target to be attacked. In NATO, for example, the job of dealing with an enemy ship in the Mediterranean has traditionally been a navy job, but if PGM are used it is not immediately apparent whether it is most efficient to employ a ship-launched, air-launched or ground-launched missile, or some combination of these. Organizationally, a task force with no bias against any of the three types would be best equipped to plan the attack. In the long run, an organization specializing in the *task* would be best suited to decide the allocation of money between the various weapon types. Similarly, the job of attacking air bases might be handled by ground-, sea- or air-launched PGM, and it might be more efficient to allocate anti-aircraft defence funds between fighters, sea-based systems and land-based systems from a single common budget.

Some complications

The practical application of PGM will naturally have its full share of complications, and using PGM will be more involved than it appears from the seven propositions above.

To begin with, the technology for accurate guidance that is most fully developed at the moment requires transmission through the atmosphere in or near the visible spectrum (see Appendix I, pp. 14–17), since simple radar guidance is not sufficiently accurate. Many present systems therefore do not work at night, or through smoke, clouds or heavy dust. Systems using long-wave infrared sensors (which will be in use by 1980) will be useful at night and will do fairly well through smoke, dust and haze, but they will be fairly expensive and may be significantly harder to maintain in the field. Nevertheless, for many years the majority of PGM will still require clear daylight.

Another problem is command and control. In past wars, commanders tens of miles behind the front concerned themselves with entire enemy divisions, or, at the very least, battalions. With PGM a division may consist of 500 separately targetable, individually moving objects. The temptation will be to handle this problem from a centralized operations room by means of data-processing technology. (There has been a trend in recent years for higher-echelon commanders to make full use of the profusion of multi-channel communications gear supplied by all-too-willing signal officers. Some senior American officers have called for an 'automated battlefield', and now some Soviet military writers are calling for a 'cybernated battlefield'.[13]) My own judgment, however, is that dealing with precision weapons will require a reversal of this trend. While it will be necessary to draw heavily on advanced data-processing techniques, especially for allocating weapons, I believe much of the solution will be found in delegation of authority and the use of standing procedures, even though the officers doing the detailed weapon control may well also be many kilometres away from the target.

A third complication is what we might call the 'sublimation' problem. If the units near the forward edge of the battle area become too small, too mobile or too well hidden to target, then the natural tendency will be to target depots and other valuable concentrations in the rear-area support structure. Thus, there is likely to be a shift to targets further and further back as longer-range PGM capable of the crucial task of finding such targets are introduced.

In the European context, this shift might find NATO at a relative disadvantage for some years, since it has been the NATO (and especially the American) style to build great depots and rely on a much larger support structure than the Warsaw Pact forces use. Quite apart from any argument for making forward forces less vulnerable, the simple fact is that, as stand-off missiles get better and more practical, action must be taken to reduce the vulnerability of rear-area concentrations – even those several hundred kilometres back and formerly thought safe from any but the most determined air attack. Like several other moves to improve preparedness for the use of PGM, this would also make NATO less vulnerable to nuclear attack, and thus help make a nuclear attack less attractive.

A further consequence of shifting attacks to targets further back will be some new attitudes towards sanctuaries. For example, the vulnerability of NATO's rear-area targets (except atomic-capable aircraft) has seldom been a major subject of concern, but now priorities must be calculated for the protection of *any* concentration

[13] See John Erickson, 'Soviet Military Power', in *Strategic Review*, Spring 1973, pp. 71 and 103–6.

of military forces or equipment targetable by stand-off weapons.

The most important complication to the simple picture is added when we consider that countering PGM will take on a very high priority (indeed, work on various kinds of countermeasures is already under way). Concealment and camouflage may work very well against present PGM, and, when they do, an attacker might logically revert to area barrage fire or to area bombing. We must therefore pause before deciding to use cities as defensive strong points. Secondly, the crews of most present PGM are vulnerable (as are airborne platforms) and will be the focus of counter-attacks. Thirdly, new designs of armour may force increases in the size of warheads – which, with shaped charges, can now be quite small.

Let us consider Proposition I again. There are complex questions of balance raised by the choice of 'many inexpensive' instead of 'fewer more expensive' vehicles. One has to ask whether the inexpensive vehicles will have the required speed, range and payload. Will the manpower required make the 'many' less desirable? Will only the 'few' be able to mount effective countermeasures devices?

There are problems, too, with the avoidance of concentrations discussed in Proposition 2. Dispersed forces may be inefficient to operate; and can an attacker's 'overwhelming superiority in a narrow sector' not be achieved by calling in offensive PGM from far away, thus concentrating the firepower but not the forces?

In fact, before my seven propositions can move out of the tentative category and become military axioms, there would need to be some force-on-force calculations of a type not carried out to date. While we can marvel at a $3,000 *TOW*-sized PGM being able to kill a $500,000 tank, we really need to calculate how many of these relatively short-range anti-tank weapons would be required on an entire front. At the same time, it would be necessary to compare a system like *TOW* with one where the individual PGM might cost more but be effective over a much wider area – for example, an RPV of 50-km range. One has to think back to the words of General Giulio Douhet, who urged the destruction of enemy bombers in the nest and not on the wing: 'How many guns [in World War I] lay waiting month after month, even years, mouths gaping at the skies on the watch for an attack which never came!' However, whether one talks about anti-aircraft or anti-tank defence, neither one-on-one calculations nor sweeping observations like General Douhet's tell the central story. This would be illuminated by force-on-force analysis, followed by considerations of how the new-style forces might affect plans and intentions as well as the interaction of forces in actual combat.

These thoughts lead naturally to some exceedingly important questions about Soviet strategy *vis-à-vis* the West and about NATO's ability to defend and deter attack. Is the present design of the Soviet Army appropriate to the task of an anti-NATO offensive? Would prudent Soviet military judgment call for a less tank-heavy posture – a ponderous move – before certifying readiness to attack?[14] These are among the most important questions to bear in mind while considering the near-term factors and those specific to NATO, which are discussed in the next two sections.

III. IMPLICATIONS UP TO 1980

It seemed useful to discuss warfare of the 1990s first, to give a sense of direction. But some quite important changes are already upon us. In this section we examine changes that will be important over the next five years, changes that are already affecting force postures and procurement decisions.

The following are some weapons developments with important consequences for our present consideration:

(1) Weapons which, though small, have effective anti-armour warheads – like the Soviet-built RPG-7 small unguided rockets used to good effect by the Egyptians in October 1973.

(2) Anti-aircraft weapons, operated by individuals or small crews, ready to use immediately after movement and cheap enough to be

[14] See Col. Edward B. Atkeson's article with the intriguing title: 'Is the Soviet Army Obsolete?', *Army*, May 1974, pp. 9 ff. (with a critical note by C. G. Jacobsen).

available in large numbers. These include several hand-held missiles, such as the Soviet SA-7 already mentioned. In the October war these weapons, along with Soviet-supplied SA-6 missiles and ZSU-23-4 guns, provided such good protection that Egyptian troops could advance without friendly air cover.

(3) Helicopters. The war in Vietnam showed the value of helicopters where opposing defences permitted. Units could be moved to difficult places without being isolated, and light payloads could be delivered tens of miles with little regard to intervening terrain. For our present purposes the point of special interest is that precision weapons are light – and so pack a great deal of capability into a helicopter-sized payload.

(4) Precision weapons for use against surface targets. These are available in great quantity: the Soviet Union supplied hundreds of *Saggers* to her Arab allies, and the United States budgeted for 30,000 *TOW* missiles and 6,000 *Maverick* in FY 1975.[15]

The important consequence of these weapons is that – setting nuclear war aside – the military balance between large-scale forces is likely to be dominated through the 1970s by a new war of numbers. The $100-million cruiser, $500-thousand tank and $10-million fighter will be challenged by the proliferation of less expensive weapons. Most are light enough to be moved easily, and many operate with almost no set-up time. There will be competition to field quantities of these relatively cheap weapons and to design them so that only modest skills are needed to operate them.[16]

However, moves to counter these abundant weapons are under way and are receiving a high priority. Spaced or array armour has been developed to counter shaped-charge warheads, in effect forcing up the weight of warhead needed to knock out a tank and, as a result, reducing the mobility of the PGM system; the United States has already designed an aircraft, the A-10, to stand up to ZSU-23-4 rounds; countermeasures against optical systems are being devised, and a whole range of tactical countermeasures developed. While we cannot know which side will be ahead in 1980, 1985 or 1990, we can see that the current measure–countermeasure contest is a new game.

Before turning to the specific case of NATO, let us consider the effects these new weapons may have on the relative position of the smaller countries. Some years ago it would have been out of the question for most small countries to install a radar network and maintain *Nike*-size missile batteries to turn back enemy air attacks; in addition, batteries in exposed locations would not have had much chance of stopping a thrust by modern armoured units. The new style of arming goes a long way towards making the small countries more defensible on both counts. For some this will mean a new set of relations between the client nation and the larger power; in other cases the small power may have substantial wealth and much independence of action, and this may have an effect on the market for munitions. Perhaps many of these countries will find it in their interest to buy more anti-tank and anti-aircraft weapons, and fewer weapons more suitable for offence. With good fortune, the net effect in many regions may be a trend towards postures that are stabilizing.

Near-term effects of precision anti-tank weapons on NATO

How might NATO use these modern weapons in the 1970s against the 15,000 or so Warsaw Pact tanks opposite its Central Front?

On the Soviet side there will be the fierce anti-aircraft defences already mentioned and quantities of at least three anti-tank missiles, *Swatter*, *Snapper* and *Sagger*, or their descendants. The Soviet Union does not seem to have an air-launched anti-tank missile.[17]

---

[15] James R. Schlesinger, *Annual Defense Department Report, FY 1975*, Department of Defense, 4 March 1974, pp. 107–8, 152.

[16] For a more complete treatment of the implications of modern military technology see Kenneth Hunt, *The Alliance and Europe: Part II: Defence with Fewer Men*, Adelphi Paper No. 98 (London: IISS, Summer 1973), pp. 14 ff.

[17] None is mentioned in 'World Missile Yearbook', *Flight International*, 8 May 1975. However, General George S. Brown has said of Soviet air-to-ground operations: 'Airborne electronic countermeasure capabilities are being upgraded. More weapons can be carried, and new weapons, including tactical air-to-surface missiles, are being developed . . . Five pylons have been observed in the *Flogger* C [a new variant with tandem seats]'. Of air-delivered precision weapons he said that, 'although our information is incomplete, we believe that we currently enjoy a considerable edge over the Soviet Union in the

On the NATO side there is a profusion of types of surface-launched missiles – with at least 16 due to be operational in the late 1970s. Nearly all these are wire-guided, and they include the previously mentioned *TOW*, *HOT* and *Swingfire*, with a maximum range of 3–4km, and the shorter-range *Dragon* and *Milan*. (For more detail, see Appendix I, particularly Table 2). Air-launched munitions include the laser-guided *Rockeye* (a cluster anti-tank munition) and *Maverick*, as well as helicopter-launched versions of *TOW*, *HOT* and *Swingfire* (for more detail, see Appendix I, Table 1).

I have already mentioned that the cost-per-round for *TOW*-class missiles is around $3,000, while the current procurement of *Maverick* has a unit cost of just under $10,000 per missile. Moreover, for automatic systems, like *TOW*, crews may be trained quickly, and there is no great problem of selection. Since, in addition, most of the systems mentioned are light and small, the number of PGM is likely to be legion. They can easily be adapted to be helicopter-mobile (though surface-launched) and should be natural candidates to serve as reinforcements or a *masse de manoeuvre*.

If all these potent properties of the new weapons are realized, it follows that there will be some new priorities on the battlefield. One of the biggest problems for all the systems mentioned above is target acquisition – though, once acquired, a target has a high probability of being destroyed unless it moves out of sight. I should therefore expect a war of seeing and hiding at the newly significant ranges of 2, 3 or 4km. If being seen at 3km leads to a high probability of being destroyed, there should be an increased use of smoke, camouflage and shielded paths for movements; equipment for night operations, and skill in using it, will also be important. This will be a competitive matter, in which the advantage goes to whichever side acquires targets at longer range.

Second, I expect both NATO and Warsaw Pact tactics will place a high priority on destroying PGM and air-defence units – by attacking either the crews or the equipment. This might be attempted by barrages of anti-personnel artillery fire, by air-dropped weapons, or by trying to take launchers under direct fire.

These considerations lead naturally to a listing of some deficiencies of the present generation of anti-tank PGM, both Soviet and NATO.

(1) Although they need to be usable at night, in bad weather and in smoke and dust, almost all current PGM depend on the visible and adjacent parts of the spectrum for guidance.
(2) Launchers and crews are relatively vulnerable to artillery barrages and scatter bomblets.
(3) The rate of fire of most PGM launchers is lower than the probable rate at which targets would appear facing them in a typical Central Front situation. Also, the time of flight of most PGM permits evasion when targets see them coming.
(4) Many PGM use small shaped charges: damage from these can be repaired, or armour redesigned to withstand them. (*Maverick* is an exception. An Israeli colonel is reported to have complained about its performance in the October 1973 war, 'The damn thing blows up those Russian tanks so much that we can't fix them up for our own use.')

The defender can do something about most of these problems. For example:

(1) Long-wave infrared systems will work well on clear nights and fairly well in dust and smoke. Scout helicopters and electronic battlefield surveillance systems will help with target acquisition.
(2) Simple means of crew protection, such as operating anti-tank missiles from under armour, should not be expensive.
(3) For the time being a defender needs a mix of guns with high-kinetic energy rounds and missiles with shaped charges. The guns have a high rate of fire close in, where seeing is less of a problem, and armour redesigned to handle shaped charges may be vulnerable to high-kinetic-energy rounds.

In sum, from the NATO point of view the PGM of the next five years – the Class of 1980 – have their potential and have their problems, but many of the latter seem to be soluble at a tolerable cost.

Supposing that most of the necessary modifications are made and that anti-tank PGM are working at somewhere near their full potential, let us

military application of this technology' (*U.S. Military Posture for FY 1976*, Statement by General George S. Brown, Chairman of the Joint Chiefs of Staff, before the Senate Armed Services Committee, Washington, February 1975, pp. 104–6).

consider a simple numerical example.[18] Let us take a NATO division facing an offensive thrust by a Warsaw Pact tank army which has 1,000 tanks and many other vehicles. Our division's task is to stop 800 of the 1,000 enemy tanks, and it has at its disposal during the first two days of conflict:

- 250 anti-tank land-based PGM launchers (*TOW*, *HOT*, *Dragon*, etc.)
- 50 PGM-equipped helicopters, capable of flying 200 sorties
- 50 fighters which can fly a total of 150 anti-vehicle sorties with 6 *Mavericks* per sortie
- Tanks, artillery, and mines sufficient to account for 100 enemy tanks.

The non-PGM weapons having stopped 100 tanks, let us assume that the total of 350 sorties by helicopters and fighters takes out another 400- (plus numerous other vehicles). Our 250 land-based PGM launchers must then stop 1·2 tanks per launcher to deal with the remaining 300 tanks (and, because the launchers would themselves be under intensive attack, these kills must be made while the defending units are still effective.

It is not the intention of this Paper to engage in detailed speculation as to the practicality of killing 1·2 tanks per land-based launcher, or as to the kill rates per helicopter or fighter sortie. However, if such a tank-killing capability *were* possible, then a very different and more hopeful picture of NATO's defensive potential will emerge in comparison with past estimates.

IV. PGM AND NUCLEAR OR CHEMICAL WARFARE

A major requirement – perhaps the major requirement – of American non-nuclear forces is that they preserve national interests without undue risk of escalation to nuclear war. For a long time there has been very little prospect of the United States or Britain embarking upon a *pre-planned* course of events leading to the use of even limited numbers of tactical nuclear weapons. There continue to be indications that the Soviet Union and France are similarly unlikely to engage in acts which involve high risks of nuclear conflict.[19] Thus, among the great dangers of military confrontation one must include situations which deteriorate rapidly, which are unexpected, and which could lead to dire consequences. This chain of events could be the more dangerous if communications were unclear, bluffs were misunderstood, or leaders were inept. Secretary Schlesinger has pointed specifically to the dangers of a general nuclear war having its origins in a deteriorating situation in Europe; the dangers would be multiplied if that situation had involved actual use of tactical nuclear weapons.

With these matters in mind, what can we say about the effect of precision-guided weapons on the probability of nuclear war? This is a subject which deserves full and separate treatment, but it is only possible here to raise a few points.

(1) The fact that defensive PGM are very potent (or could be made so) could go a long way towards stopping a tank thrust without resorting to tactical nuclear weapons. The potency of PGM and RPV in attacking targets in rear areas will be quite high by the 1980s and may substitute for medium-range nuclear strikes.

(2) On the other hand, a single nuclear warhead in the kiloton range might damage 100 vehicles dispersed over several square kilometres; to cause equal damage with present PGM might require 100 successful non-nuclear weapons. Non-nuclear area weapons are possible, but their technology is not yet well-developed, and they might be quite heavy. However, in addition to providing area coverage, tactical nuclear weapons have also been considered to compensate for inaccurate guidance and poor target location, and for this PGM do provide an adequate substitute. Some types, like RPV, can even home in on targets which have moved.

A more complete consideration of this point

[18] The following example is very much a creature of its assumptions, and so is *not* a forecast.

[19] See the chapter by Thomas W. Wolfe in Kurt London, ed., *The Soviet Impact on World Politics: A Symposium* (New York: Hawthorn, 1974). Also Hannes Adomeit, *Soviet Risk-Taking and Crisis Behaviour: From Confrontation to Coexistence?* Adelphi Paper No. 101 (London: IISS, Autumn 1973). On the attitudes of French leaders see Marc Ullmann, 'Security Aspects in French Foreign Policy', *Survival*, November/December 1973. This discussion stresses the French emphasis on the political values of their nuclear forces. But French officials talk freely of early nuclear use in a deteriorating situation.

would compare the effectiveness of precise non-nuclear weapons and nuclear weapons, with effects tailored for each of the several categories of target: dispersed and soft, dispersed and moderately hard (like tanks), hard point targets, targets of imprecisely known location, etc. From a purely technical point of view, some targets could be most effectively attacked with nuclear weapons and some with non-nuclear weapons.

(3) Some Western writers have considered the use of precision guidance to permit effective use of sub-kiloton nuclear warheads – so-called 'mini-nukes'. In part, these weapons were advocated because it was believed that the American strategic nuclear threat has become so decoupled from events in Europe that any American tactical nuclear contribution should be designed to repel an attack at its outset, and that early release of small yields would be more credible and result in less collateral damage. However, critics have questioned the physical effectiveness and nuclear efficiency of sub-kiloton weapons and have been concerned about the blurring of the nuclear firebreak.[20] If both sub-kiloton weapons and highly effective non-nuclear weapons were available the question for NATO would not be: Which can do the military job? It would be: Which (by possession or use) gives the signal most consistent with NATO goals?

(4) A strategy which aims at terminating or de-escalating a conflict will profit from the execution of precise, predictable, and understandable combat operations. The criteria for damage must be that damage to intended targets should be maximized, and to non-targets minimized. Non-nuclear PGM meet these requirements and in a great many cases serve a conflict-limiting strategy well. On the other hand, the consequences of even the most limited use of nuclear weapons are quite unknowable, and this makes their use risky if the goal is to limit conflict.[21]

(5) We discuss below how modern weapons are likely to increase the rate of destruction in non-nuclear war, as well as the rate and cost of munitions consumption. This faster rate could lead to surprises, or to a pause where there might be some temptation to escalate to nuclear use. However, if we have anticipated the pause, and especially if we have been observing limitations in the preceding fighting, the pause could lead to de-escalation and negotiation.

Some Soviet strategic writings seem to regard non-nuclear war as a phase preceding nuclear war. If non-nuclear PGM do, in fact, halt a Warsaw Pact tank thrust, how will the Soviet Union calculate the value of going nuclear at that time? Two points seem straightforward. First, operating in small, separated units is a good tactic for PGM warfare and foı avoiding tactical nuclear vulnerabilities. Second, passive protection and dispersal of rear-area support facilities is a good idea in any case. And to diminish both kinds of vulnerability is to decrease the other side's incentive to use nuclear weapons.

But the most important point is that the technology for precision delivery has come at just the time when Western strategy is turning towards the threat of carefully controlled combat – both in regional and in intercontinental conflict – as a more credible deterrent than the threat of unrestrained response.

Another topic which can only be mentioned here is the way the introduction of chemical warfare might affect these conclusions about PGM defences. There have been a number of warning signals in Soviet writings that they take chemical war preparations seriously. The Soviet-supplied chemical warfare equipment carried by captured Arab troops during the October 1973 war led to a public statement by the US Defense Department on the need for defensive preparations. At the least, this should make Western force planners look favourably on PGM systems that can be operated from enclosed vehicles or bunkers.[22]

[20] On 23 May 1974, the United States made a statement to the Geneva Disarmament Conference which 'gave assurance . . . that it would not develop a new generation of miniaturized weapons on the battlefield'. Dr Fred C. Ikle, director of the US Arms Control and Disarmament Agency, said in an interview: 'We have no intention to move in a direction that could blur the distinction between nuclear and conventional arms' (*New York Times*, 24 May 1974).

[21] See Albert Wohlstetter, 'Threats and Promises of Peace: Europe and America in the New Era', *Orbis*, Winter 1974, pp. 1122 ff.

[22] For a more complete discussion, see General George S. Brown's statement in *U.S. Military Posture for FY 1976*, *op. cit.* in note 17, pp. 114–120.

V. CONCLUSION

Subject to the several qualifications and omissions which have been pointed out, three main conclusions emerge.

First, the advent of PGM is probably advantageous to the defender. Target acquisition is the key to their successful use, and it is much easier for a defender to hide than for his opponent, who is moving through unfamiliar terrain and has no opportunity to prepare positions. Moreover, PGM, being relatively light, can be moved quickly to where they are needed – perhaps by helicopter – while heavier systems, including tanks, might arrive too late. Current PGM are not well suited to attack, since most of them are designed for specific defensive tasks; hopefully, therefore, the acquisition of PGM by both sides will lead to a more stable situation.

However, it is necessary to be cautious about concluding definitively that these developments always favour the defender (some years from now new target-acquisition systems, longer-range PGM and area munitions may help an attacker). Moreover – taking the NATO case again – even a massive Warsaw Pact offensive may involve the attacker holding defensively along 95 per cent of the front, while thrusting offensively along the other 5 per cent. And NATO forces must, in many places, go locally on the offensive. Weapon trends which assist efficient holding, therefore, will also suit Warsaw Pact purposes in the majority of places. The problem is to overcome NATO's capability to hold defensively in the remaining few places – where great concentrations of Warsaw Pact offensive strength would be needed. Again we come back to the question that requires detailed study: just how vulnerable to PGM are such concentrations?

A second conclusion is that an important consequence of the dispersal of so much destructive power down to small units, and the natural delegation of authority to use it, will be that the pace of war will be faster. In places with large concentrations of forces there will be an unprecedented intensity of non-nuclear conflict. Even though, as noted earlier, the total weight of munitions to do a job may decrease over the entire time of the conflict, the rate of use – in terms of proportion of stocks consumed – is likely to increase. The material destroyed is likely to be something like ten times greater than we have been thinking about for non-nuclear war. We had a glimpse of this in the sudden logistic demands of the October 1973 war, but a war in Europe could dwarf those consumption rates. Will this pace lead to escalation or negotiation, when forces find munitions and equipment largely spent after three or four days?

Thirdly, there is a hopeful sign that the trend of the first part of this century, towards the inclusion of non-military target systems and civilian populations in military campaigns, will be reversed. Precision delivery means that military targets can be destroyed with less total explosive power and less collateral damage to non-military targets. The faster pace discussed above means that tactical forces-in-being, as well as strategic forces, count for more, and the general economy for less, in achieving a favourable outcome.

It is interesting to hypothesize that, given precision weapons, it may now be possible, with little loss of military efficiency, to adhere to a rule which strictly limits civil damage in both offensive and defensive operations. I should like to see this hypothesis explored in a broad analysis, and – if the result is favourable – it could be an appropriate subject for international discussion and negotiations leading to an agreed limitation.

Finally, another prospect for an arms-limitation agreement should be considered: one which seeks stability through an emphasis on defensive capability. Expensive, large, multi-purpose weapons (tanks, fighter bombers such as the F-14, nuclear aircraft carriers) are usually well-suited to offence. To the extent that smaller PGM-equipped units are making such systems less viable, it could perhaps be demonstrated that both sides would be served by limiting the numbers of such large systems. For a given budget or manpower ceiling, more resources could go into defensive units that would hold so well against an attack that, a generation from now, service school graduates may quote a new maxim: 'The best defence is a good defence'.

I must end on a note of caution. More precise long-distance delivery and the packing of more destructive potential into small packages is resulting in the rearrangement of traditional categories of weapons and systems (Proposition 7, page 5). At the same time there is increased interest in limited and carefully controlled operations, even at intercontinental range. For example, in the American forces, the long-range bombers of Strategic Air Command may be adapted to launch precision non-nuclear weapons while the aircraft of other commands may be armed with new weapons which make them more interchangeable with the 'strategic' bombers, except in range and payload. Arms-control agreements relating to 'strategic forces' will need to be replaced by agreements more directly related to the physical capabilities of delivery vehicles and their payloads, supplemented by agreements on the restrained use of forces.

Agreements on vehicles or substantial physical systems cannot be vitiated in the short-term, nor be responsive to short-term stimuli. But those which call for restraint in operations must stand up to the harder test of still seeming mutually desirable under the many short-term stimuli in the haze of war.

Appendix I

Current PGM and Guidance Techniques

Tables 1, 2 and 3 list some of the more significant PGM for use against land targets and aircraft; many of the weapon types which have caused the current wave of interest in PGM are included, but not all the types listed have been in large-scale production.*

A glance at Tables 1, 2 and 3 shows that a large number of systems in widespread use depend on the transmission of light in the visible spectrum (or, in the case of those which use lasers, just outside it). Thus many systems do not work well at night or in adverse weather (clouds, fog or haze), and their range could be limited by battlefield smoke or the deliberate use of obscurants. On the other hand, when conditions permit, visible-spectrum systems provide good resolution (and high angular accuracy), can use many inexpensive and widely available aids (such as telescopes, prisms, and television equipment), and present pictures in a familiar form for human operators to recognize and understand. The closer such a picture is to real-life scenes, the easier a man can use all his well-developed powers of visual observation. In fact, almost all current target acquisition for PGM involves a man looking at a scene or a pictorial representation of a scene, recognizing a chosen target among other things, placing a cross-hair or cursor over it, and pulling a trigger or actuating an automatic seeker. However, countering visual recognition is a familiar military task, and may not be very expensive. Camouflage, smoke screens, visual decoys, etc., can all be effective.

After launch the PGM may be guided by a homing device (for example, one which directs a missile towards the hottest spot in its field of view) or by commands from the launch point (the simplest systems fly the munition as if it were a model aircraft). A third type of guidance directs the munition (normally a missile) to a set of grid co-ordinates, which may be map co-ordinates or temporary electronic co-ordinates. This last type, which is usually not sufficiently accurate to qualify as 'precision' guidance, is not noted in Tables 1–3 but will be discussed in the section on mid-course guidance on page 23.

Particularly where fixed targets are involved, many countermeasures can be overcome by using area correlators (also called map-matchers) which compare a reference picture taken during previous reconnaissance, perhaps from considerable altitude, with the current picture seen from the missile. Since it will be impractical for a defender to change all the shapes recorded on the reference picture, the usual countermeasures will not work. This kind of guidance can give high accuracy, even though the missile may have travelled a long way and its launch position may not be precisely known. On the other hand, the prior reconnaissance mission does have to survive, afford recognition and produce a usable reference map. Correlators can be used with infrared or microwave sensors also, and for the latter their anti-jamming properties are even more necessary. A simpler system of the same sort can be used when the target itself is camouflaged or hard to recognize: the missile can be guided with respect to an offset aim-point. This requires that the attacker know both his own and the target's distance and azimuth from some easy-to-identify terrain feature.

However, the main impetus to use the non-visual spectrum comes from the need to operate in darkness and in bad weather. The options available will be discussed in Appendix II.

* For a brief treatment of almost all current guided missiles, see 'World Missile Yearbook' in *Flight International*, 8 May 1975; for a more extensive treatment, with many diagrams and photographs, see R. T. Pretty and D. H. R. Archer (eds), *Jane's Weapon Systems, 1974–1975* (London: Jane's Yearbooks 1974). For the technically trained, *Aviation Week* gives frequent detailed reports on progress in development and production of PGM systems, with emphasis on engineering choices, production problems and funding.

Table 1: Selected Air-to-Surface PGM for the late 1970s

Designation	Developed by	Range (km)	Weight of round (lb)[a]	Guidance	Comments
AS.20	France	7	315	Radio command	
AS.30	France	12	1,150	Radio command, some automatic features	Can be used against hard points and ships
AS.37 *Martel*	France/ Britain	30–60	1,170	Passive radio frequency seeker	For attacks on radars
AJ.168 *Martel*	Britain/ France	30–60	1,215	TV link + radio command	Can be steered by landmarks till target comes into TV camera range
Maverick AGM-65A	USA	[b]	460	TV tracker	Tracks automatically after lock-on
Laser-guided *Rockeye* KMU-420	USA	[c]	500	Laser homing	Primarily for anti-tank
Modular-guided Glide Bomb	USA	80(?)	2,000	TV or other + radio command	Evolving programme related to MK84-*HOBOS*
Shrike AGM-45A	USA	12–16	390	Passive radio frequency seeker	Used against ground radars
Standard ARM AGM-78A/B/C/D	USA	725	1,795	Passive radio frequency seeker	Used against ground radars
MK82-LGB KMU-388/B	USA	[c]	500	Laser homing	Steerable bomb
MK84-LGB KMU-351/B	USA	[c]	2,000	Laser homing	Steerable bomb
MK84-*HOBOS* KMU-353A/B	USA	[c]	2,000	TV homing	Steerable bomb
MK84-IR *Paveway* KMU-359/B	USA	[c]	2,000	Infrared homing	Steerable bomb
Bulldog AGM-83A	USA	10	598	Laser homing	Derived from older *Bullpup*
Condor AGM-53A	USA	110	2,110	Electro-optical/ TV homing or command	Remotely piloted
Walleye I	USA	[c]	1,100	Electro-optical/ TV homing	Carried by F-4, A-7, etc.
Walleye II	USA	[c]	2,330	Electro-optical/ TV homing or command	Used against large, semi-hard targets: e.g., bridges, ships

[a] In this table only, to agree with common American and British usage, weights are given in pounds. Bomb weights are nominal and actual weights may be 10 per cent more or less.

[b] Aerodynamic range has been given as 22km, but practical range for guided launch has not been released.

[c] Free fall.

SOURCES: *Flight International*, 14 March 1974; *Jane's Weapon Systems, 1973–1974; Aviation Week*, 10 December 1973, pp. 13 ff. and 15 July 1974, pp. 265 ff.

Table 2: Selected Surface-to-surface PGM for late-1970s Anti-armour Use

Designation	Developed by	Min./max. range (m)	Weight of round (kg)	Guidance	Comments
ENTAC	France	400/2,000	12	Manual command	Production complete, 13,000 produced
SS-11/AS-11	France	350/3,000	29·9	Manual command	Anti-submarine version for helicopters, 160,000 produced
SS-12/AS-12	France	800/6,000	76	Semi-automatic	Anti-submarine version for helicopters 8,000m max. range
HOT	France/ Germany	75/4,000	22	Semi-automatic command	May be used in Bö-105 light helicopter
Cobra	Germany	400/2,000	10·3	Manual command	No connection with the American AH-1Q *Cobra* helicopter
Milan	France/ Germany	25/2,000	6·7	Semi-automatic command	Two-man crew
Swingfire	Britain	< 150/4,000	34	Manual command + aids	Operator can be offset 100m, vehicle in defilade
Shillelagh MGM-51C	USA	(?)	27	Semi-automatic command, infrared link	Fired from 152mm gun mounted on the M-551 *Sheridan* or M-60A2 medium tank
TOW BGM-71A	USA	65/3,750	19	Semi-automatic command	Many carriers, but M-113 APC and AH-1Q *Cobra* predominate in US Army. (See note under *Cobra* above.)
Dragon M-47	USA	?/1,000	6·35	Semi-automatic command	Man-portable
Snapper AT-1	USSR	500/2,300	22	Manual command	Used in several Pact armies. Mounts on BRDM armoured reconnaissance vehicle.
Swatter AT-2	USSR	(?)	about 20	Manual command, infrared terminal guidance	Mounts on APC and BRDM reconnaissance vehicle.
Sagger AT-3	USSR	500/3,000	11	Manual command	Mounts on APC and BRDM reconnaissance vehicle (which carries 6 under a retractable plate).

SOURCES: *Flight International*, 14 March 1974; *Jane's Weapon Systems, 1973–1974.*

Table 3: Selected Surface-to-air PGM for the late 1970s

Designation	Developed by	Altitude limits (m)	Guidance	Comments
Crotale	France	50/3,000	Multi-mode	
Roland	France/ Germany	15/3,000	Optical (*Roland* I) radar or optical (*Roland* II)	Mounts on *Marder* or AMX-30; adopted by US Army in January 1975
Rapier	Britain	5,000 (?)	Optical or with *Blindfire* radar	Mounts on two Land Rovers
Blowpipe	Britain	(?)	Optical + remote control	Shoulder-fired
Redeye MIM-43A	USA	(?)	Optical aiming infrared homing	Shoulder-fired
Chaparral MIM-72A	USA	(?)	Optical then infrared	Modified *Sidewinder* 1C, typically used with *Vulcan* gun
Stinger XFIM-92A	USA	(?)	Optical aiming, infrared homing	Shoulder-fired, in development to replace *Redeye*
SA-6 *Gainful*	USSR	about 18,000	Radar or optical	Mounted in threes on a tracked transporter
SA-7 *Grail*	USSR	50/1,500	Optical aiming, infrared homing	Shoulder-fired or in batteries on trucks

SOURCE: *Flight International*, 14 March 1974.

Appendix II
Wavelength Effects

A variety of factors influence the choice of wavelength to be used in a guidance system. Here we will discuss what is given up (especially in accuracy) and what is gained (especially in seeing through obscurants, including murky weather) by guidance systems not using the visible spectrum (usually defined as the wavelengths from 0·4 to 0·7 microns*). Again, this is a discussion for the general reader rather than the technical specialist.

The parts of the electromagentic spectrum most often used for self-contained guidance systems are:

Infrared:	around 0·9 microns, 3 to 5 microns, and 8 to 14 microns.
Millimetre wave:	around 8·6mm (35 GHz), 3·2mm (94 GHz), and 2·1mm (140 GHz)
Other microwave:	3cm (10 GHz), 6cm (5 GHz), and 10cm (3 GHz).

A rough idea of the loss in accuracy as wavelength goes up can be gained by considering what happens to angular resolution (the smallest angle between two equal point targets at which the viewing device can tell there are two targets, not one). Accuracy generally varies with resolution, but also depends on a complex set of other factors – for example, the signal-to-noise ratio and how long a radar beam stays on the target. Angular resolution is generally given by

$$R = \frac{\lambda}{D} k$$

where R is the resolution in radians
λ is the wavelength in centimetres
D is the aperture or antenna diameter in centimetres
k is a constant depending on system design and operation

For a given D, therefore, the angular resolution increases (i.e., deteriorates) proportional to wavelength. Thus, for example, a 3cm radar with a 75cm antenna diameter can typically get angular resolutions of 2° to 3°, but a 30cm radar would need a 750cm antenna. By contrast, the eye can resolve angles to about a minute, or 0·017°. Practical results with good visual bombsights show that cross-hairs can be set to within about 0·06°.

Infrared systems will be treated in the most detail below, since they can be used in a wide variety of terminal-guidance devices and are likely to be in widespread use by 1980.†

First, however, let us consider how often non-visual systems would be needed, on the basis of some data on central European weather.

Surface visibility and ceiling in central Europe

Central European weather, particularly in the region from the Ruhr to Stuttgart, is characterized by ground fog, industrial haze and frequent low cloud ceilings. R. E. Huschke has analysed a large body of data, taken over a number of years at several places in central Germany, and some of his results are shown in Figures 1 and 2.

Using data from standard visibility measurements made by observers looking out to see if they can distinguish familiar large landmarks, Figure 1 shows how the range at which a moving tank can be detected varies with surface visibility. Assuming the tank is continuously visible, using H. H. Bailey's target acquisition model,‡ the two curves show detection range both for the unaided eye and for an observer with 6-power binoculars. The bars at the bottom of the figure show how often one can expect various surface visibilities; for example, winter visibility would be at or under about 2·5km a quarter of the time, and under 5km half of the time. Note that for these two examples the practical detection ranges of just above and just below a kilometre are much less than the 3km maximum range of several anti-tank PGM.

The curves in Figure 2 show how often air-delivered munitions can be used under two different assumptions as to ceiling and visibility; they also show the extent of daylight hours. These curves, devised by

---

* One micron, increasingly called a 'micrometre', is one-millionth of a metre. In the radio-frequency bands frequencies in Gigahertz (GHz = 10^9 cycles per second) are often used instead of wavelengths. The approximate relationship between the two is

$$\lambda f = 30$$

where λ is the wavelength in centimetres
f is the frequency in GHz.

† The remainder of this appendix draws on material originally prepared by S. J. Dudzinsky, C. C. Chen, L. G. Mundie, R. E. Huschke and P. A. CoNine, all of The Rand Corporation.

‡ H. H. Bailey, *Target Detection through Visual Recognition: A Quantitative Model* (Santa Monica, Calif.: Rand Corporation, RM-6158/1-PR, 7 February 1970).

Figure 1: Relationship of detection range to visibility in central Germany in daylight hours

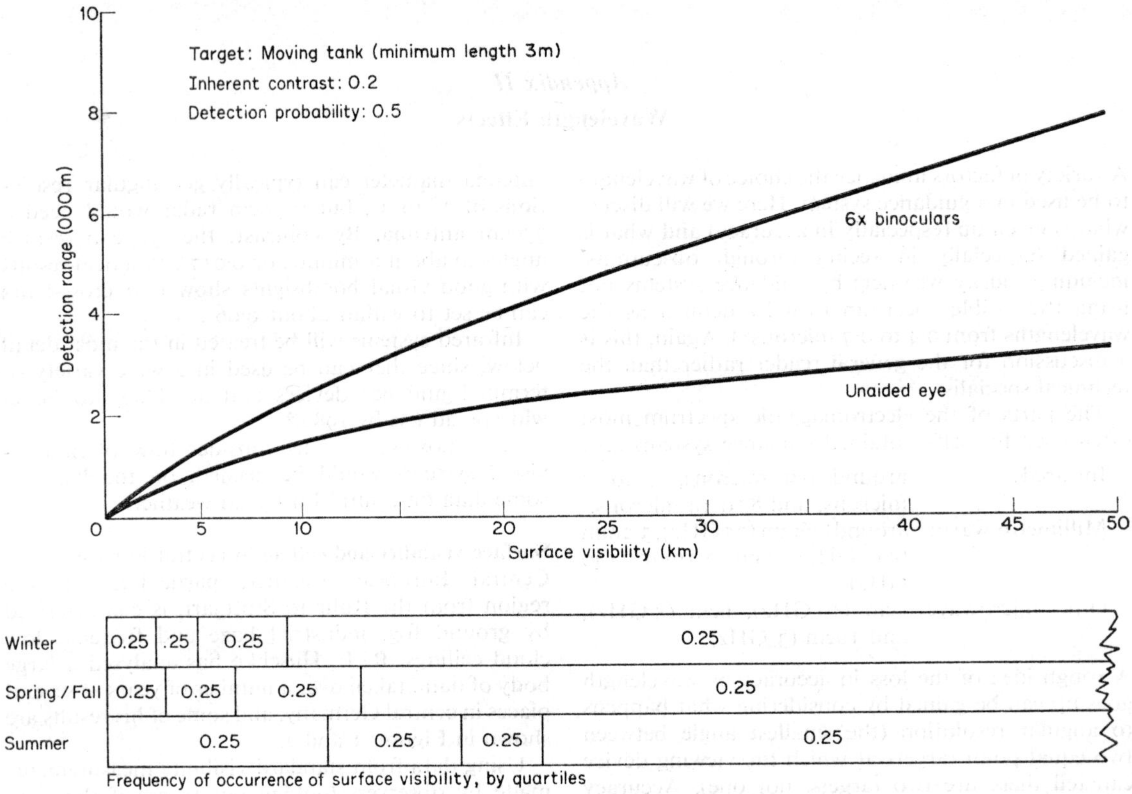

Figure 2: Probability that ceiling and visibility values meet given limits near Berlin

(a) Ceiling 10,000 ft or more and visibility 4 miles or more

(b) Ceiling 1,000 ft or more and visibility 3 miles or more

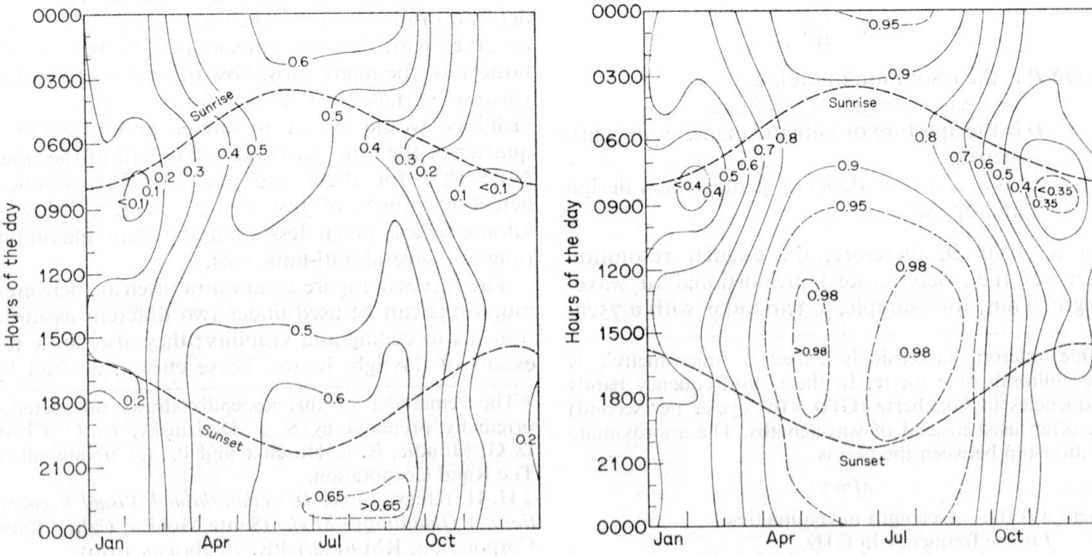

Figure 3: Transmittance over a 6,000-ft horizontal path at sea level in a clear atmosphere containing 10 g/m³ precipitable water

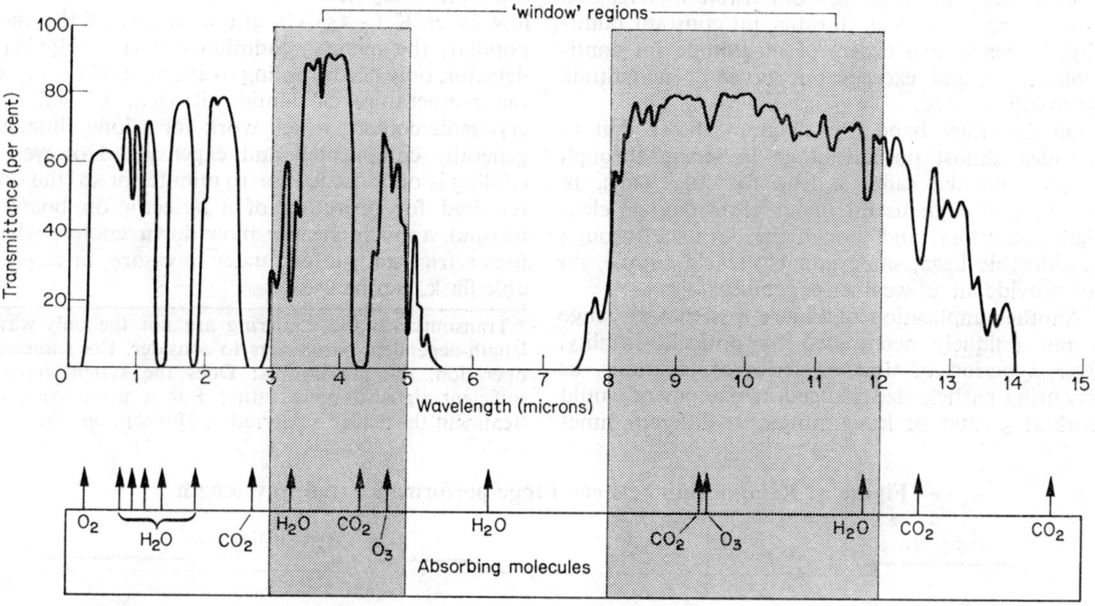

SOURCE: Adapted from R. D. Hudson, Jr., *Infrared System Engineering* (New York: John Wiley, 1969).

Huschke, are a bit difficult to follow at first glance, but they do provide a great deal of useful information. Figure 2(b), for example, shows that on a July afternoon the combination of 1,000-ft ceiling and 3-mile visibility can be expected between 95 and over 98 per cent of the time. At sunrise in December, however, it could be expected only about 40 per cent of the time, and there would be 16 hours of darkness.

Infrared systems for PGM

The infrared (IR) portion of the electromagnetic spectrum includes wavelengths from about 0·7 microns to more than 100 microns. Detectors designed for use above 3 microns often use the apparent temperature difference between target and background to get a useful signal. Thus, they can be passive, homing on the target's own emissions of radiation, and can detect targets under night-time as well as daytime conditions. Although such detectors could be developed for use in any portion of the IR spectrum, most effort has been devoted to development in the 3–5 micron and 8–14 micron 'window' regions, where attenuation of emitted energy due to atmospheric absorption is at a minimum.

The transmittance (proportion of energy transmitted) at the relevant wavelength over 6,000ft in clear atmosphere (i.e. no fog or haze particles) is shown in Figure 3. However, attenuation by water vapour absorption can be significant even in the 'window' regions, and one can see from the diagram that, assuming the water vapour content of the atmosphere is 10 grammes per cubic metre, the average transmittance over the 8–12 micron region is about 70 per cent. (The water content mentioned corresponds to 60 per cent relative humidity at 20°C.) With a similar clear atmosphere containing less than 5 gm/m³ of water vapour (typical of mid-latitude winter conditions) and over 15 gm/m³ (representing warm, moist summer conditions) average transmittance would be about 85 per cent and about 55 per cent respectively.

One must add to the effect of water vapour the effect of scattering by particles in the atmosphere (e.g. by haze and smoke). If the diameter of particles is less than about one-tenth of the IR wavelength, then such attenuation is almost negligible (typical haze particles range from about 0·5 micron to under 1 micron, and phosphorous smoke particles are about the same size). However, fog and cloud (which are water particles) do pose problems for IR systems, because their sizes are comparable to IR wavelengths. (Attenuation by rain is not as severe, since raindrops are much sparser than cloud and fog particles.)

The effect of scattering on the detection range is shown in Figure 4 which, taking a particular type of cloud (stratus) and two types of haze, shows the relative value of 'generalized meteorological range' (i.e., detection range) as a function of wavelength. One can see that, in going from visible wavelengths to 10 microns, the gain in range (at constant humidity) is nearly two orders of magnitude for continental haze, and exceeds one order of magnitude for maritime haze.

On the other hand, the diagram shows that IR provides almost no advantage in seeing through clouds, and the same is true for fog. Thus, IR detectors, though useful under clear day or clear night conditions, and though they can see through considerable haze, smog and battlefield smoke, do not provide an all-weather capability.

Another implication of Figure 4 is that IR range is not uniquely determined by optical visibility. Over a period of time or series of locations, as obscuring particle size changed, IR equipment would work at greater or lesser ranges at different times when the optical visibility was the same.*

To maximize detectivity, most IR detectors available today need cryogenic cooling: some to as low as 20°K (−253°C), although one of the most popular, the mercury-cadmium-telluride (HgCdTe) detector, only needs cooling to about 80°K (−193°C) the temperature of liquid nitrogen. Closed-cycle cryogenic coolers, which work for a long time, are generally complicated and expensive. However, if cooling is only needed for 10 minutes or so (the time required for operation of a detector on board a missile), a much cheaper blow-down cooler, which uses refrigerant stored under pressure in a refillable flask, can be used.

* Transmittance and scattering are not the only wavelength-dependent parameters to consider. For successful operation, the question is: Does the system have a sufficient signal-to-noise ratio? For a more complete treatment the reader is referred to Hudson, *op. cit.*

Figure 4: Relationship between range performance and wavelength

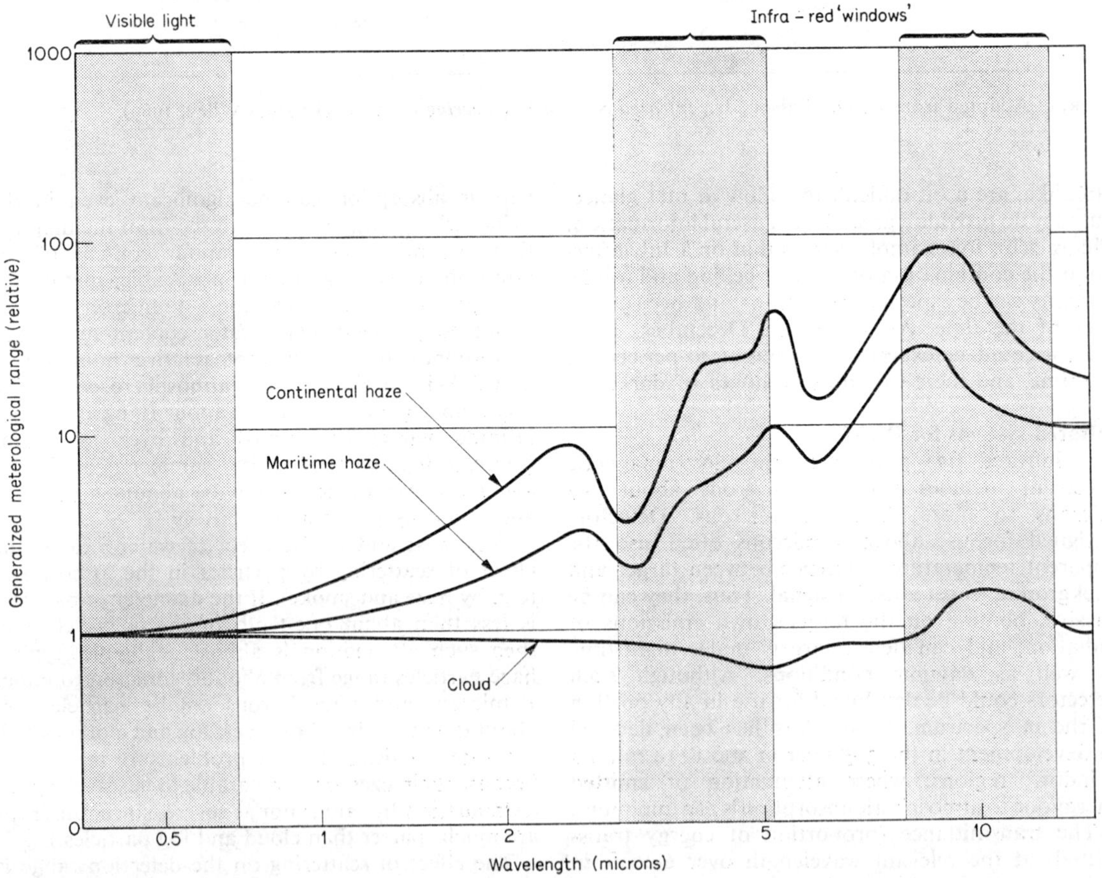

SOURCE: Adapted from D. Deirmendjian, *Electromagnetic Scattering on Spherical Polydispersions* (New York: American Elsevier, 1969) by R. E. Huschke.

Millimetre wave radiometry

One might expect electromagnetic sensors in the millimetre wave region to be useful in guiding PGM, since, by the equation on p. 18, modest antenna diameters would give excellent resolution by comparison with traditional microwave radars. One interesting kind of equipment to explore would be a radiometer (that is, a passive receiver) which would use directional antennas and superheterodyne receivers to detect thermal emission in the microwave region (though at these frequencies the observed signals obtained in field trials arise primarily from the emissivity differences between various objects, rather than from true temperature differences).

Microwave radiation penetrates clouds and fog reasonably well, and this is the main reason for seriously considering it in this application. Microwave radiometers are passive (hence their operation is covert, and their power requirement low), and they are relatively inexpensive and rugged. Their ability to generate good terrain pictures, even through fairly thick clouds, makes them useful for navigation and for attacks against fixed, pre-briefed targets, in conjunction with either manual control or area correlators. However, there are difficulties in using them for detecting and attacking moving targets.

The main weaknesses of practical passive microwave radiometers are the poor ratio of target signal to background clutter signal usually obtainable and the fact that they are relatively easy to jam. Figure 5 shows the probability of detecting a truck at a slant range of 3·6km for one set of reasonable parameter choices (including the assumption that the difference in emissivity between the truck and a field of grass which forms its background is about 0·9). At the assumed frequency of 35 GHz fog would have little effect over such a short transmission path; however, one can see that the antenna diameter needed for a 90 per cent detection probability exceeds 6 metres – much too unwieldy for most PGM applications.

The performance of radiometers operating at the two frequently used higher frequencies of 94 and 140 GHz is better than that of the 35-GHz radiometer in clear weather, since the antenna diameter required is inversely proportional to the frequency of operation. In certain adverse weather conditions, however, the increased atmospheric attenuation, the decreasing sensitivity of receiver circuits as frequency increases and other technical factors might combine to prevent satisfactory operation at these higher frequencies. Thus, millimetre-wave radiometers may not be satisfactory in detecting targets like trucks. They would still be useful, however, for navigation and for attacking fixed targets whose location in relation to major terrain features is known. Major terrain features would always be recognizable; the clutter problem does not apply, and obscuring them by jamming could be difficult.

Other microwave radar

Most guidance applications of microwave radar in the 'X' (3cm) or 'S' (10cm) bands are for anti-ship or anti-aircraft uses. Practical antenna sizes do not give the angular resolution needed to attack tanks

Figure 5: Probability of detecting truck with passive microwave radiometer at 35 GHz

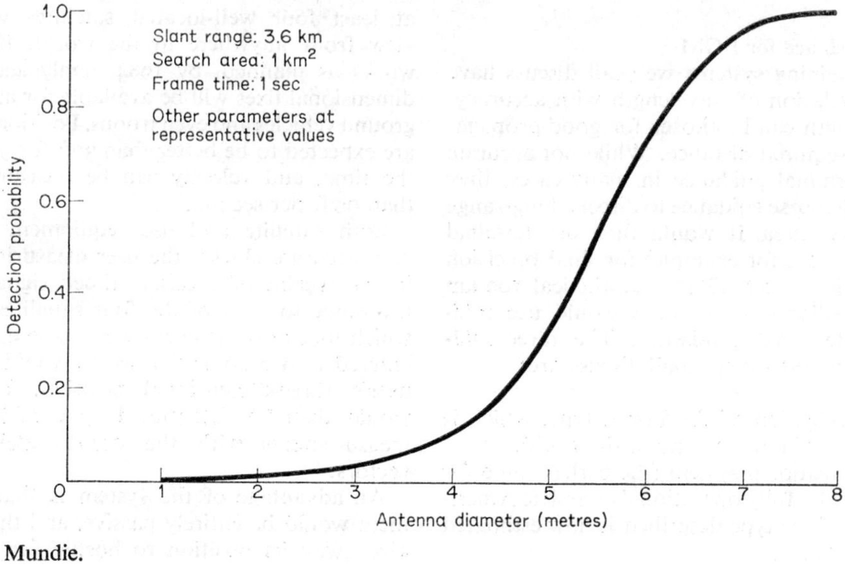

SOURCE: L. G. Mundie.

or trrucks. On the other hand, with ships and aircraft, not only is there great contrast between the target and its background but the target is usually surrounded by empty space. Anti-ship missiles tend, for a variety of reasons, to be large enough to take antennas of 30cm diameter or more; a simple radio altimeter can be used for guidance in the vertical plane; and the antenna beam need only scan around a single axis. For anti-aircraft use, the contrast of aeroplane against sky is even greater, while the missile is usually small, very manoeuvrable in response to guidance signals and is able to use a proximity fuse.

'X' and 'S' band radars are either active (with transmitter and receiver carried on the missile) or semi-active (the transmitter staying near the launch point to illuminate the target for an on-board receiver). Semi-active systems have advantages in keeping missile-weight low and in removing the transmitter's directly radiated 'spill-over' energy from the vicinity of the receiver. The latter is particularly useful where clutter from the ground must be filtered away in order to reveal echoes from the target, which is often done by using the doppler frequency shift. This shift, which depends on the velocity of the target, makes it possible to sort out and display only the echoes from fast-moving aircraft.

The accuracy with which most such systems can hit a target will depend on the pattern of radar returns reflected from the target, the aerodynamic characteristics of the missile, and the guidance accuracy of the radar. The dominant component of miss distance may often be due to the apparent shifting of the centroid of the echo as the relative position of target and radar changes.

Mid-course guidance for PGM

The three remaining systems we shall discuss have no strong correlation of wavelength with accuracy, so the wavelength can be chosen for good propagation over the required distance. While not accurate enough for terminal guidance in many cases, they could give mid-course guidance to direct a long-range PGM to a target area. It would then use terminal guidance (IR or TV, for example) for final precision guidance to the target. (The hypothetical 100-km cruise missile discussed on p. 3 would use mid-course, then terminal, guidance). The three mid-course guidance systems we shall discuss are:

Hyperbolic navigation of the Loran type, which is available today in many parts of the world.

A satellite navigational system (the earliest time for deployment of a fully operational complete American system of the type described is now estimated to be about 1984).

The combination of time-of-arrival and distance-measuring equipment systems, recently developed for attacks on missile guidance radars.

There are two basic modes of mid-course navigation: (1) an autonomous mode, in which the PGM receives navigation signals and processes them on board to determine its position and the required navigational corrections, and (2) a co-operative mode, in which the PGM retransmits the navigation signals it receives to a separate station where its position and navigational corrections are computed. Hyperbolic and satellite systems can use either.

In addition to the radio-frequency systems mentioned here, an inertial guidance device is sometimes used for mid-course guidance or for spatial reference. Inertial systems also often play a role in giving initial orientation to PGM, for example, as part of their airborne fire control systems.

Loran is one of several so-called 'hyperbolic' navigation systems. Precisely timed pulses are transmitted by two pairs of stations many kilometres apart. Lines of equal difference in the arrival time of the signal from one pair at the missile's receiver are hyperbolas, and points in space can be located by plotting the time differences from both pairs on a map over which the two families of hyperbolas are printed as a grid. Systems have been developed which automatically guide a missile to a point in the hyperbolic co-ordinate system, and this is accurate enough for mid-course guidance.

The US Department of Defense is currently developing a satellite navigation system: NAVSTAR Global Positioning System (GPS). The plan calls for three rings of eight satellites each in 12-hour circular inclined orbits, a configuration which ensures at least four well-located satellites would be in view from anywhere in the world. If the system works as planned, by 1984 highly accurate three-dimensional fixes will be available for aircraft, ships, ground vehicles and even troops. Positional accuracies are expected to be better than 30ft for 90 per cent of the time, and velocity can be measured to better than 0·2ft per second.

Both satellite and user equipment would have very accurate clocks, the user measuring the delays in the receipt of precisely timed signals, and thus the range to each of the four satellites and rate at which that range is changing. These data would be entered into a computer which would calculate the user's three-dimensional position. User velocity would then be determined by combining doppler measurements with the user-to-satellite position vectors.

An advantage of the system is that user equipment would be entirely passive, and thus would not give away its position to hostile forces; it is also said to contain a number of anti-jamming features.* The smallest of the user sets could be carried in back-packs by individual soldiers and is estimated to cost around $16,000, while 'Spartan' sets for a few thousand dollars, are a possibility. The user equipment for mid-course missile guidance would be similar but probably a bit more complex.

Time-of-arrival/Distance-Measuring Equipment (TOA/DME) was originally conceived by the US Air Force as a combination to detect and locate a radiating missile-guidance radar (using the TOA part) and then guide a munition to it (using the DME part). The TOA part of the system, designed to fix the source of radiation quickly (in case it is shut down to avoid further detection), locates active radio targets by noting the differences in the time of arrival of the same signal at two or more airborne stations whose separation could be measured by DME. The DME-guided weapon, which might be a glide bomb, is then launched. It carries a relatively inexpensive transponder (a transmitter which pulses when an incoming pulse is received) which responds to signals from each of two base stations. The round trip time of each signal and response from each base station to weapon and back is used to calculate range to weapon. This gives a circle of position from each base station, and the intersection of the two circles gives the location (the ambiguity caused by the fact that the circles intersect in two places can usually be resolved). DME can be used without TOA for guiding a weapon to a target located by other means. It is said to have an accuracy of about 100ft, enough to damage a soft target if the warhead is large. However, this kind of system might guide the weapon to the target area, when a terminal seeker (more accurate, but capable only of short-range use), perhaps using IR, would home on the target itself.†

* *Aviation Week*, 15 April 1974, pp. 22 ff.

† For further details, status reports, and discussions of combining various techniques see *Aviation Week*, 10 December 1973, pp. 13 ff. and 15 July 1974, pp. 265 ff.

ADELPHI PAPERS

The following is a selection of those available. They may be ordered from the Institute at a current price of **25p (75c)** *per copy, post free. The current price of Nos* 70, 71, 79, 90–112 is **35p ($1)**, *the current price of No* .113 *onwards is* **50p ($1.50)** *post free.*

No. 61. British Defence Policy: The Long Recessional by L. W. Martin. November 1969.
No. 62. Military Logistic Systems in NATO: The Goal of Integration. Part I: Economic Aspects by Geoffrey Ashcroft. December 1969.
No. 64. Survey of Strategic Studies. January 1970.
No. 65. Soviet–American Relations and World Order: Arms Limitations and Policy. Papers from the Scheveningen Conference. February 1970.
No. 66. Soviet–American Relations and World Order: The Two and the Many. Papers from the Scheveningen Conference. March 1970.
No. 69. Ethics and Deterrence: A Nuclear Balance Without Hostage Cities? by Arthur Lee Burns. July 1970.
No. 70. Europe and America in the 1970s: Part I: Between Détente and Confrontation. Papers from the Evian Conference. November 1970.
No. 71. Europe and America in the 1970s: Part II: Society and Power. Papers from the Evian Conference. November 1970.
No. 73. Wary Partners: The Soviet Union and Arab Socialism by Robin Buss. December 1970.
No. 78. Future Conditional: The Prospect for Anglo-French Nuclear Co-operation by Ian Smart. July 1971.
No. 79. Urban Guerrilla Warfare by Robert Moss. August 1971.
No. 80. Moving the Glacier: The Two Koreas and the Powers by Morton Abramowitz. September 1971.
No. 82. Civil Violence and the International System: Part I: The Scope of Civil Violence. Papers from the Stresa Conference. December 1971.
No. 84. Mutual Force Reductions in Europe: The Political Aspects by Christoph Bertram. January 1972.
No. 86. Nuclear Safeguards by Ryukichi Imai. March 1972.
No. 90. Japanese Economic Policies and Security by Richard Ellingworth. October 1972.
No. 91. East Asia and the World System: Part I: The Super-Powers and the Context. Papers from the Ste. Adèle (Quebec) Conference. November 1972.
No. 92. East Asia and the World System: Part II: The Regional Powers. Papers from the Ste. Adèle (Quebec) Conference. November 1972.
No. 93. Conflicts in Africa. December 1972.
No. 94. The US Forces and the Zero Draft by Morris Janowitz. January 1973.
No. 95. Japanese Security and the United States by Kunio Muraoka. February 1973.
No. 96. The Alliance and Europe: Part I: Crisis Stability in Europe and Tactical Nuclear Weapons by Wolfgang Heisenberg, Summer 1973.
No. 97. Options for Japan's Foreign Policy by Masataka Kosaka. Summer 1973.
No. 98. The Alliance and Europe: Part II: Defence with Fewer Men by Kenneth Hunt. Summer 1973.
No. 99. Nuclear Weapons and Chinese Policy by Harry Gelber. Summer 1973.
No. 100. Insurgency and Counter-Insurgency in Rhodesia, 1957–1973: An Account and Assessment by Anthony R. Wilkinson. Autumn 1973.
No. 101. Soviet Risk-taking and Crisis Behaviour by Hannes Adomeit. Autumn 1973.
No. 102. Force in Modern Societies: Its Place in International Politics. Papers from the Travemünde Conference. Winter 1973.
No. 103. Force in Modern Societies: The Military Profession. Papers from the Travemünde Conference. Winter 1973.
No. 104. Indonesia's Future and South-East Asia by Peter Polomka. Spring 1974.
No. 105. New Approaches to Arms Reduction in Europe by J. I. Coffey. Autumn 1974.
No. 106. Nuclear Forces for Medium Powers: The Strategic Requirements: Part I: Targets, Weapons Systems and Deterrence by Geoffrey Kemp. Autumn 1974.
No. 107. Nuclear Forces for Medium Powers: Parts II and III: Strategic Requirements and Options by Geoffrey Kemp. Autumn 1974.
No. 108. The Alliance and Europe: Part III: Weapons Procurement in Europe – Capabilities and Choices by Roger Facer. Winter 1974.
No. 109. The Alliance and Europe: Part IV: Military Doctrine and Technology by Steven Canby. Winter 1974/75.
No. 110. American Foreign Policy in the Nixon Era by Anthony Hartley. Winter 1974/75.
No. 111. The Arab–Israel War, October 1973 – Background and Events by Elizabeth Monroe and A. H. Farrar-Hockley. Winter 1974/75.
No. 112. Defence Budgeting: The British and American Cases by Richard Burt. Winter 1974/75.
No. 113. Prospects for Nuclear Proliferation by John Maddox. Spring 1975.
No. 114. The Middle East and the International System: Part I: The Impact of the 1973 War. Papers from the Sussex Conference. Spring 1975.
No. 115. The Middle East and the International System: Part II: Security and the Energy Crisis. Papers from the Sussex Conference. Spring 1975.
No. 116. Strategic Deterrence Reconsidered by
No. 117. Oil and Influence:

Forthcom

Military Vincent.

Discount r